Table of Contents

1 Introduction to Seismic Design

Earthquakes are unpredictable, powerful forces of nature. They can cause catastrophic damage to buildings, leading to loss of life and significant economic consequences. Understanding the fundamental principles of seismic design is, therefore, paramount for ensuring the safety and resilience of structures in earthquake-prone regions. This understanding starts with recognizing that seismic waves, originating from the earth's movement along fault lines, propagate through the ground, causing vibrations that affect building foundations and structures. These vibrations, characterized by their frequency and amplitude, dictate the type and extent of damage a structure may sustain.

Designing for seismic loads differs significantly from designing for static loads. Static loads, such as the weight of a building, are constant and predictable. Seismic loads, however, are dynamic and vary in intensity and direction. This dynamic nature necessitates the use of specialized analytical techniques to accurately assess structural response during an earthquake. Ignoring these dynamic effects can lead to severe underestimation of the forces acting on a building, potentially resulting in structural failure.

Building codes and regulations play a vital role in ensuring seismic safety. These codes, developed by organizations like the International Code Council (ICC) and other regional bodies, specify minimum design requirements for new buildings and, increasingly, for retrofitting existing ones. These regulations are based on extensive research and experience and aim to balance safety with economic feasibility. Codes incorporate factors such as the seismic hazard level of a region, the type of soil conditions, and the occupancy of the building. Adherence to these codes is essential to obtain building permits and ensure that buildings are designed and constructed to withstand earthquakes to a certain standard. They are regularly updated as research progresses and more knowledge about earthquake behaviour is acquired. Understanding these codes is not just about compliance, it's about understanding the underlying principles which drive the safety requirements.

The design process typically involves several steps, beginning with a thorough site investigation to understand the local soil conditions and seismic hazard. This is critical because soil characteristics significantly influence the ground motion experienced during an earthquake. Subsequently, the structural engineer uses various analytical models to estimate the forces acting on the building during an earthquake scenario. These models are sophisticated and take into account various factors, including the building's geometry, material properties, and the characteristics of the earthquake itself. The outcome of this analysis determines the required strength and ductility of the structural elements to ensure the building can withstand the anticipated seismic loads.

A key concept in seismic design is ductility. Ductility refers to a material's ability to deform significantly before failure. This is crucial in earthquakes because ductile structures can absorb energy during ground shaking, preventing catastrophic collapse. Brittle materials, on the other hand, tend to fail abruptly, leading to sudden and potentially widespread damage. Therefore, the selection of materials with appropriate ductility is vital for seismic resistance. The design process frequently involves the selection of structural systems and materials specifically chosen for their ductility.

Beyond the initial design stage, regular inspections and maintenance are essential for maintaining the seismic resilience of buildings over time. Building deterioration, whether due to aging, environmental factors, or poor maintenance, can significantly impact a building's ability to withstand an earthquake. Regular checks help to identify potential weaknesses and address them proactively. This preventative approach is significantly more cost-effective than undertaking emergency repairs following an earthquake. Further, proactive maintenance reduces the risk of sudden, catastrophic failure and subsequent loss of life. Regular inspection programs are important to prevent future damage which could weaken the structure's resilience to earthquake activity.

In conclusion, mastering the fundamentals of seismic design is a crucial step in creating earthquake-resilient structures. It demands a comprehensive understanding of seismic principles, building codes, and advanced analytical methods. By combining theoretical knowledge with practical applications, engineers can design and retrofit buildings to minimize damage and ensure the safety of occupants during seismic events. This knowledge forms the bedrock for the subsequent chapters, which delve deeper into specific aspects of seismic hazard assessment, structural analysis techniques, retrofitting methodologies, and material selection, building on this foundation to provide a thorough understanding of the field. The significance of integrating this understanding into every stage of a building's lifecycle, from design to maintenance, cannot be overstated.

1.1 Basic Seismic Principles

Earthquakes are sudden, violent ground movements caused by the release of energy along fault lines within the Earth's crust. This energy propagates as seismic waves, which travel through the ground, causing the shaking that damages structures. Understanding the nature of these waves - their different types (P-waves, S-waves, surface waves), their velocities, and how they interact with the ground and buildings - is fundamental to seismic design and retrofitting. The frequency content of these waves is crucial; higher frequencies tend to cause more localized damage, while lower frequencies affect the entire structure.

Seismic waves interact with buildings in complex ways, depending on the building's size, shape, and material properties. A building's natural frequency, or the rate at which it vibrates freely, is a key parameter. If the frequency of the incoming seismic waves matches the building's natural frequency, resonance occurs, leading to amplified vibrations and potentially catastrophic failure. This phenomenon is a primary concern in seismic design, requiring careful consideration of building characteristics to avoid resonance. Understanding this interaction helps engineers design structures that can withstand the dynamic forces of an earthquake. Properly understanding these fundamentals allows engineers to make informed decisions.

Designing earthquake-resistant structures involves minimizing these harmful effects. This is achieved through various strategies, including increasing a building's stiffness and ductility. Stiffness refers to a structure's resistance to deformation, while ductility is its ability to deform significantly without fracturing. A stiffer building will resist movement more effectively, but excessive stiffness can lead to brittle failure. Conversely, a more ductile structure can absorb energy more effectively during an earthquake, minimizing damage. Finding the optimal balance between stiffness and ductility is a key challenge in seismic engineering.

Building codes and regulations play a crucial role in ensuring seismic safety. These codes set minimum standards for seismic design, requiring engineers to account for earthquake hazards in their designs. Regulations specify acceptable levels of risk and define the design requirements needed to achieve those levels. Codes also provide guidelines on appropriate materials, construction techniques, and inspection procedures. These requirements evolve as our understanding of seismic behavior and engineering materials improves. Adherence to these codes is critical for protecting lives and minimizing economic losses.

The severity of ground shaking varies considerably depending on factors like the magnitude of the earthquake, the distance from the epicenter, and the local soil conditions. Soft soils tend to amplify ground motion, increasing the seismic hazard in those areas. Understanding these site-specific conditions is therefore crucial in seismic design. Engineers use ground motion prediction equations and site-response analyses to estimate the potential ground shaking at a given location. This information, which incorporates probability-based approaches to account for uncertainties, is essential in the decision-making process. These assessments are never perfect, and often include some degree of uncertainty.

Seismic analysis involves calculating the forces that an earthquake will impose on a structure. This analysis considers various factors, including the type of ground motion expected, the building's mass and stiffness, and the damping capacity of structural components. Advanced analytical techniques, like nonlinear dynamic analysis and pushover analysis, simulate the building's behavior during a severe earthquake, providing insights into its potential response. These analyses are critical for assessing the structural integrity and performance of both new and existing buildings.

Finally, remember that seismic design is an iterative process. Engineers often use computer models to simulate the effects of earthquakes on buildings. These models allow for refinement of the design to meet the required safety standards. Through this cyclical approach, a safe and effective design can be developed, improving structural resilience. These models can even predict the behavior of different designs to minimize the impact of seismic events. The goal is always to ensure the structure can withstand the expected ground shaking, even in the most challenging seismic environments. This iterative process is fundamental to good seismic engineering practice.

1.2 Building Codes & Regulations

Building codes and regulations are the bedrock of seismic safety. They're not just a collection of rules; they represent decades of research, countless case studies, and a deep understanding of how structures behave during earthquakes. Think of them as a carefully constructed framework, designed to ensure that buildings can withstand the forces of nature, protecting lives and investments. Compliance isn't optional; it's a legal mandate, safeguarding the public and minimizing potential losses. The specific requirements vary significantly based on geographical location, the type of construction, and the building's intended use.

Navigating this complex landscape can be daunting for beginners. However, understanding the core principles will empower you to make informed decisions. One crucial aspect is the determination of seismic zones. These zones are classified based on historical earthquake activity and geological data, reflecting the probability and intensity of future seismic events. Buildings in high-risk zones face stricter regulations, demanding higher standards of design and construction. This often translates into reinforced structures, specialized foundations, and the incorporation of damping systems.

Then there are the design requirements themselves. Building codes specify minimum standards for structural elements, such as load-bearing walls, columns, beams, and foundations. These standards aren't arbitrary; they're based on detailed calculations and simulations designed to ensure adequate strength and ductility - the ability of a structure to deform without fracturing - under seismic loads. This includes considerations for the materials used. Different materials behave differently under seismic stress; concrete, steel, masonry

- each has its own strengths and weaknesses.

Furthermore, codes often dictate specific detailing requirements. These may involve aspects like the proper placement of reinforcement bars within concrete elements, the use of specific connection types for beams and columns, or the inclusion of shear walls to enhance lateral resistance. These seemingly minor details can have a dramatic impact on a building's overall performance during an earthquake. Neglecting these details can lead to catastrophic failures, even in structures that appear robust on the surface. You should always consult with experienced structural engineers who are fully aware of the local codes and the best practices for seismic design.

Regular inspections and maintenance are also vital. Codes often incorporate requirements for periodic inspections of structures, particularly those in high-risk areas or older buildings. These inspections are designed to detect any signs of deterioration or damage, allowing for timely repairs and preventing larger problems from developing. Maintenance is about more than just aesthetics; it's about preserving the structural integrity of a building and ensuring its continued ability to withstand seismic events. Ignoring this critical component compromises the seismic resilience.

Finally, understand that building codes are living documents. They're constantly updated to reflect new research, technological advancements, and lessons learned from past earthquakes. Staying abreast of these changes is crucial for ensuring the ongoing safety and compliance of any seismic structure. Regularly consulting with industry experts and keeping informed about revisions to the codes are essential to upholding the highest standards of seismic safety. This continuous learning and adaptation ensures that our structures are continually evolving to meet the challenges posed by these powerful natural events. Understanding these codes is not merely about following rules; it's about protecting lives and securing investments in a world increasingly affected by seismic hazards. It demands respect for the forces of nature and a commitment to building a safer future.

2 Seismic Hazards & Risk Assessment

Ground motion prediction is paramount in seismic hazard assessment. We use various techniques, including probabilistic seismic hazard analysis (PSHA), which considers the likelihood of different earthquake magnitudes and distances, coupled with ground motion prediction equations (GMPEs) to estimate the likely shaking intensity at a specific site. This involves identifying active faults, understanding their recurrence intervals, and characterizing the soil conditions. These factors significantly influence the amplification of seismic waves, making accurate site-specific analysis essential for informed decision-making in seismic design and retrofitting projects. Failure to accurately account for these variations can lead to significant underestimation of seismic hazards and subsequent inadequate design provisions. Ignoring the specific geological characteristics can result in catastrophic failures during actual seismic events. Therefore, a thorough understanding of local geology and accurate prediction of ground motion are critical for effective seismic risk mitigation.

The accuracy of the models used, and the data employed are just as important as the methodology itself.

Vulnerability analysis helps quantify the potential damage to a building during an earthquake. We examine a building's structural characteristics, material properties, and design features, using methods such as fragility curves, which probabilistically link the intensity of shaking to the likelihood of different levels of damage. This involves a detailed inspection of the building, including the condition of its foundations, walls, and connections. Non-destructive testing methods, such as ground penetrating radar or ultrasonic testing, can supplement visual inspections, providing valuable information about the internal state of the structure. Specific attention should be paid to weak points in the structure, such as unreinforced masonry walls or inadequate connections between structural elements. The results of these analyses provide critical input for developing cost-effective retrofitting strategies. Accurate vulnerability assessment is not merely a theoretical exercise; it directly informs the prioritization of buildings requiring retrofitting, which allows for the most effective use of limited resources.

Effective risk mitigation requires a multifaceted approach. It's about understanding the potential losses resulting from both direct physical damage and indirect consequences like business interruption. We develop seismic risk maps to illustrate areas with varying levels of hazard. These maps combine ground motion predictions with vulnerability assessments, providing a spatial visualization of the potential impact of an earthquake. These risk maps are crucial in guiding land use planning, building code development, and the prioritization of retrofitting projects. Ultimately, the objective is to strike a balance between the cost of implementing mitigation measures and the potential economic and social costs of earthquake damage. This balance demands a sophisticated understanding of the trade-offs involved and requires a collaborative approach amongst engineers, policymakers, and the community at large. Decisions made without such a comprehensive view can result in insufficient protection or unnecessary expenditures. The importance of integrating diverse perspectives into the decision-making process cannot be overstated.

2.1 Ground Motion Prediction

Accurate ground motion prediction is paramount for effective seismic design and retrofitting. We can't reinforce buildings effectively without understanding the forces they'll face. This requires a multifaceted approach, integrating various data sources and sophisticated modeling techniques. Ignoring this crucial step can lead to inadequate designs and compromised safety, ultimately resulting in unnecessary damage or even collapse during a seismic event.

Probabilistic seismic hazard analysis (PSHA) forms the cornerstone of modern ground motion prediction. This method doesn't just predict a single ground motion scenario; rather, it provides a range of potential ground motions and their associated probabilities. This probabilistic approach is far more realistic than deterministic methods which only consider a single, often worst-case, scenario. It helps engineers make informed decisions about acceptable risk levels. The information is incorporated into the design process to ensure buildings have a high probability of surviving various seismic intensities. One of the key inputs for PSHA is the seismic source characterization. This involves identifying all potential earthquake sources within a region, estimating their recurrence rates, and defining their magnitudes. Geological surveys, historical earthquake records, and paleoseismological studies all contribute to this critical process. Furthermore, understanding fault characteristics, like rupture length and slip rate, is vital for precise ground motion estimation. The accuracy of the source characterization directly influences the accuracy of the predicted ground motions, highlighting the critical nature of this initial stage.

Ground motion prediction equations (GMPEs) are empirical relationships that link earthquake magnitude, distance from the source, and site conditions to the predicted ground motion parameters. These parameters typically include peak ground acceleration (PGA), peak ground velocity (PGV), and spectral accelerations (SA) at different periods. Numerous GMPEs exist, each developed based on different datasets and methodologies. Selecting the appropriate GMPE is crucial and depends on the specific regional geology and seismicity.

Care must be taken to select GMPEs appropriate for the local conditions to avoid inaccuracies in the estimations. The selection requires experience and judgment, considering factors such as the proximity to active faults, soil type, and the nature of the seismic waves.

Site characterization plays a significant role in refining ground motion predictions. Local soil conditions can amplify or attenuate seismic waves, leading to significant variations in ground motion intensity across even relatively short distances. Geotechnical investigations, including borehole measurements and seismic surveys, are crucial for determining the soil profile at the site. This information allows for accurate input into ground motion prediction models. For instance, soft soils tend to amplify ground shaking, while stiff rocks tend to reduce it. The complexity of different sites' soil composition requires specialized knowledge in geotechnical engineering for effective characterization.

Modern ground motion prediction incorporates advanced computational techniques. Numerical simulations, such as finite-element or finite-difference methods, can model the propagation of seismic waves through complex geological structures. These simulations can account for three-dimensional variations in subsurface properties and improve the accuracy of ground motion predictions, particularly in complex geological settings. These sophisticated modeling techniques are frequently employed to determine detailed characteristics of seismic wave behavior for accurate risk assessment.

In conclusion, accurate ground motion prediction is a complex but critical step in seismic design and retrofitting. It involves a detailed understanding of seismic sources, careful selection of ground motion prediction equations, thorough site characterization, and the application of sophisticated computational techniques. By integrating all these aspects, engineers can develop more realistic and robust assessments of seismic hazards and ensure that buildings are designed and retrofitted to withstand the forces of earthquakes. The accuracy of these predictions directly impacts the safety and resilience of structures in earthquake-prone regions. Neglecting this essential step can lead to substandard designs and endanger lives and property.

2.2 Vulnerability Analysis

Understanding a building's vulnerability to seismic events is paramount for effective retrofitting. This involves a thorough evaluation of its structural components, materials, and overall design against anticipated seismic loads. A crucial first step is meticulously documenting the existing building's condition, including detailed architectural drawings, material specifications, and any previous repairs or modifications. This forms the foundation upon which subsequent analyses will be built. Without this comprehensive documentation, accurate vulnerability assessment becomes impossible. Next comes a careful visual inspection of the structure. Look for signs of distress like cracks in walls, columns, or foundations; evidence of previous settlements; or deterioration of materials. Pay close attention to the building's connection points, where different structural elements meet. These are often weak points vulnerable to damage during an earthquake. Photography and detailed notes are essential at this stage to record all findings accurately. Consider engaging a structural engineer experienced in seismic assessments; their expertise is invaluable in identifying subtle yet significant issues that might be missed by untrained eyes.

Further analysis should incorporate advanced techniques. For instance, nonlinear dynamic analysis can simulate the building's response to various earthquake intensities and durations. This involves using sophisticated software to model the building's structural behavior, accounting for the nonlinear behavior of materials under stress. Such analyses can predict potential damage patterns and identify critical areas needing strengthening. Don't underestimate the power of a properly executed nonlinear analysis, the results will guide your retrofitting strategy more efficiently than relying solely on visual inspection alone.

The capacity spectrum method offers a more efficient approach for preliminary assessment. By comparing the building's capacity (its ability to resist seismic forces) with the anticipated demand (forces from earthquakes), this method provides a rapid evaluation of seismic performance. While less detailed than nonlinear analysis, it's a valuable tool for screening buildings and prioritizing those needing immediate attention. Remember that the capacity spectrum method is a simplified representation, it should be complemented by more detailed analysis for critical structures.

Your analysis should also include a thorough assessment of the soil conditions beneath the building. Poor soil conditions can significantly amplify ground shaking, leading to increased structural damage. Geotechnical investigations, including soil sampling and laboratory testing, are often necessary to determine the soil's properties and its potential impact on the building's seismic performance. Failing to account for soil characteristics can render your retrofitting strategy ineffective and potentially dangerous.

Finally, integrate the findings from all these assessments. Consider each component's vulnerability: the foundation, columns, beams, walls, connections, and the overall structural system. This comprehensive analysis will pinpoint the building's weaknesses, highlighting areas requiring strengthening or other retrofitting interventions. Remember that this isn't just about identifying problems; it's about understanding how these weaknesses will likely behave during an earthquake, the potential extent of the damage, and how to mitigate that damage effectively through retrofitting. This integrated approach ensures that the chosen retrofit strategy addresses the building's specific vulnerabilities. This holistic understanding of the building's behavior is the key to developing a robust and effective retrofitting plan. This detailed analysis, combining visual inspection, advanced analytical techniques, and soil investigations, is vital in ensuring the building's long-term safety and resilience against future seismic events. The resulting vulnerability assessment serves as a crucial roadmap for guiding the subsequent retrofitting process.

2.3 Risk Mitigation Strategies

Identifying and addressing potential weaknesses is paramount. A thorough visual inspection, often overlooked, can reveal crucial details like cracks in masonry, damaged foundations, or corroded steel elements. This initial assessment informs subsequent steps, directing resources towards the most critical areas. Ignoring this foundational step can lead to costly oversights and inadequate mitigation. Remember, a picture is worth a thousand words, especially when documenting pre-retrofit conditions. Detailed photographic records are essential for insurance claims and future reference.

Next, consider employing advanced non-destructive testing (NDT) methods. Ground Penetrating Radar (GPR) can unveil subsurface anomalies impacting foundation stability. Ultrasonic pulse velocity (UPV) testing assesses the integrity of concrete elements, identifying areas weakened by age or deterioration. These techniques provide objective data far surpassing visual inspection alone, allowing for precise identification of vulnerable zones and a more accurate assessment of the overall structural health. Investing in NDT is a strategic investment in long-term safety and cost-effectiveness.

Focusing on specific structural elements is key. Masonry walls, a common weakness in older buildings, often require specialized attention. Techniques such as adding steel jacketing, strengthening existing mortar, or introducing shear walls can dramatically improve their seismic performance. For foundations, soil improvement techniques like dynamic compaction or ground grouting may be necessary to increase bearing capacity and prevent settlement. The choice of mitigation technique should always be tailored to the specific building, considering its age, materials, and existing structural defects. This personalized approach is crucial for optimal results.

Beyond structural elements, consider the influence of non-structural components. Unsecured heavy objects, such as equipment or ceiling fixtures, can cause significant damage and injuries during an earthquake. Securing these items through proper anchoring or bracing is a cost-effective measure significantly enhancing occupant safety and preventing secondary damage. This often overlooked aspect is crucial for minimizing the overall impact of an earthquake. Don't undervalue the importance of seemingly insignificant details; they can collectively contribute to substantial damage.

Furthermore, regular maintenance and inspection schedules are crucial. A proactive approach is far more effective and cost-efficient than reactive repairs after an earthquake. Regular inspections, ideally conducted annually, allow for early identification of minor issues before they escalate into major problems. This preventative approach also extends the lifespan of the building and reduces the overall risk of structural failure. Implementing a clear maintenance plan and adhering to it strictly is a crucial element of successful risk mitigation. This commitment demonstrates responsible building management.

Finally, the integration of seismic isolation systems should not be dismissed. These systems, strategically placed between the foundation and the superstructure, effectively decouple the building from the ground motion. This dramatically reduces the seismic forces transmitted to the structure, significantly enhancing its resilience. Though initially more expensive, the long-term benefits in terms of reduced repair costs and life-cycle cost savings, often justify the upfront investment. Proper consideration of seismic isolation demonstrates a commitment to long-term protection. The value of safeguarding lives and infrastructure far outweighs initial costs. Choosing the right system and experienced installer are critical factors in implementation. Selecting the wrong isolation system can lead to ineffective mitigation. Therefore, thorough research is essential before any decision is made.

Remember, effective risk mitigation isn't a one-size-fits-all solution. A comprehensive strategy requires a thorough understanding of the building's specific vulnerabilities and a tailored approach to addressing them. This involves a detailed assessment, the selection of appropriate techniques, and a commitment to ongoing maintenance. The goal is not simply to meet minimum code requirements, but to enhance the building's resilience and protect its occupants. A robust risk mitigation strategy contributes significantly to a building's longevity and the safety of its inhabitants, showcasing the value of thoughtful planning and responsible action. The investment in seismic resilience is not an expense but a profound commitment to long-term safety and sustainability.

3 Structural Analysis for Seismic Loads

Understanding how structures respond to seismic forces is paramount. Linear analysis, a simplified approach, assumes a structure's behavior remains elastic even under significant shaking. This method, while computationally efficient, often underestimates the damage potential, particularly for older buildings or those constructed with materials prone to brittle failure. Therefore, its use should be carefully considered, especially when dealing with structures that have already experienced damage or those designed according to older, less stringent codes. Relying solely on linear analysis could lead to a false sense of security, ultimately jeopardizing the safety of occupants.

Nonlinear analysis provides a more realistic depiction of structural behavior. It accounts for the material's nonlinear characteristics, accurately capturing the complex interplay of forces during an earthquake. This approach is critical when evaluating the structure's performance at high-intensity ground motions. It involves sophisticated modeling that incorporates factors such as material degradation and yielding, enabling a more accurate prediction of structural response and potential collapse mechanisms. The computational demands are significantly greater than linear analysis; however, the increased accuracy makes it indispensable for performance-based seismic design.

The Capacity Spectrum Method (CSM) offers an efficient way to assess a structure's seismic performance. It efficiently combines the structure's capacity (its ability to resist forces) with the demands imposed by the earthquake (the forces it experiences). By plotting both on a single graph - the capacity and demand spectra - engineers can quickly identify potential vulnerabilities. This allows for a targeted approach to retrofitting, focusing resources on the areas most susceptible to damage. Essentially, CSM allows for the optimization of design modifications, balancing cost-effectiveness with seismic safety.

Pushover analysis, a nonlinear static procedure, simulates the gradual increase of lateral forces on a structure. The analysis reveals how the structure responds as these forces grow, pinpointing weak points before complete failure. This allows engineers to visualize the progression of damage, identifying potential collapse mechanisms. Visualizing this helps to optimize the strengthening process. The results directly inform reinforcement strategies, ensuring efficient allocation of resources to significantly enhance seismic resilience.

However, it's important to remember that pushover analysis relies on simplified assumptions; therefore, it's usually paired with other analysis methods to get a more complete picture of structural behavior.

Performance-based design aims to achieve a pre-defined level of performance during and after an earthquake. Instead of simply meeting minimum code requirements, this approach focuses on defining acceptable damage limits for various earthquake intensities. This enables engineers to design or retrofit structures to withstand specific seismic events with predictable outcomes. This approach prioritizes functional recovery after the earthquake and is increasingly recognized as a more holistic and responsible approach to seismic design, especially in areas with high seismic risk. It needs thorough understanding of performance objectives, detailed modeling, and careful consideration of risk factors.

Advanced analytical techniques extend beyond the basic methods. These may include finite element analysis (FEA), which provides a detailed representation of the structural components and their interactions. Furthermore, these methods include sophisticated constitutive models that accurately capture the behavior of materials under cyclic loading conditions. They may also involve the use of advanced numerical techniques such as higher-order elements and adaptive meshing for improved accuracy, especially in complex structural systems. These analyses require considerable computational resources and expertise; hence, they are often used in conjunction with simpler methods to validate or refine design decisions, focusing primarily on complex or critical structural elements and systems. The use of these techniques helps engineers to achieve a high degree of confidence in the building's ability to withstand severe seismic events.

3.1 Linear & Nonlinear Analysis

Linear analysis simplifies the complex behavior of structures under seismic loading by assuming a proportional relationship between load and displacement. This approach, while computationally efficient, often underestimates the true response, especially for structures exhibiting significant nonlinearity, such as those nearing collapse. We use simplified models that assume the structure's material properties remain constant throughout the earthquake's duration. These models are useful for preliminary design checks and initial assessments, particularly when dealing with simple structures or in the early stages of a project. They provide a fundamental understanding of the building's response and serve as a solid foundation for more sophisticated analyses.

Nonlinear analysis, in contrast, directly addresses this limitation by explicitly modeling the material's nonlinear behavior and its effects on the structure's overall response. This is particularly crucial for older structures where material degradation might be present, significantly altering the structure's behavior under seismic loads. It accounts for factors like cracking, yielding, and even member failure that linear analysis ignores. This level of detail enables a more realistic prediction of the building's performance during an earthquake, providing a far more accurate picture of potential damage.

The transition from linear to nonlinear analysis is not arbitrary; it depends greatly on the complexity of the building and the level of accuracy required. Simple structures might suffice with linear analysis, while complex multi-story buildings with irregular geometries definitely necessitate a nonlinear approach. Factors like the presence of weak stories, significant material degradation, or unusual architectural details all point toward the need for nonlinear analysis to gain a more realistic understanding of the seismic response. The choice is a careful balance between computational cost and the desired accuracy.

Nonlinear analyses can also incorporate different levels of complexity. Some methods will only account for material nonlinearities, while others consider geometric nonlinearities, which incorporate the changes in a structure's geometry as it deforms under load. These intricate models offer significant benefits for structures subjected to severe earthquakes but are more computationally demanding. Therefore, careful consideration of the project's specific requirements and constraints— including budget and time limitations— should be made before opting for this approach.

A crucial aspect of nonlinear analysis involves selecting appropriate material models that accurately capture the behavior of the various components of the structure. Concrete, for example, exhibits complex nonlinear behavior under cyclic loading, which must be accounted for by using advanced constitutive models. Similarly, the behavior of steel, especially after yielding, requires careful consideration. The use of incorrect material models can lead to inaccurate predictions and compromise the integrity of the analysis, potentially resulting in inadequate seismic design or retrofitting strategies. Consequently, validating the chosen models against experimental data is a critical step in ensuring accuracy.

Effective nonlinear analysis often involves the use of sophisticated numerical techniques, such as finite element analysis (FEA). FEA breaks down the structure into a multitude of small elements, each with its own material properties and behavior. The interaction between these elements is analyzed numerically under various loading conditions, simulating the earthquake's effect on the structure. The outputs of these analyses provide comprehensive insight into the behavior of the building during an earthquake, including the location and severity of potential damage. The precision achieved from sophisticated techniques ensures that the design recommendations are aligned with the realities of seismic conditions.

The results from nonlinear analysis can then be used to guide the design or retrofitting process. This information is critical in making informed decisions about structural modifications to enhance seismic resilience. For example, if the analysis highlights a weak story, reinforcing this section of the building becomes a priority. Similarly, any significant material deficiencies uncovered during the analysis would necessitate modifications to ensure adequate structural strength. The objective of both linear and nonlinear analyses, when applied correctly, is to prevent collapse during a seismic event and minimize damage.

Choosing between linear and nonlinear analysis is a critical decision in seismic design. While linear analysis provides a simplified, initial assessment, nonlinear analysis offers a more realistic representation of the building's response under strong seismic loading. Selecting the appropriate level of analysis depends on numerous factors, including the structure's complexity, the anticipated severity of the seismic hazard, and the level of detail needed in the analysis. The ultimate goal is to create a structure capable of withstanding seismic events while minimizing the impact on its functionality and safety. This necessitates a careful consideration of the relative strengths and limitations of each analytical approach. Therefore, consulting structural engineers with relevant expertise is a necessary prerequisite.

3.2 Capacity Spectrum Method

The Capacity Spectrum Method (CSM) offers a powerful, yet relatively straightforward approach to evaluating the seismic performance of structures. Unlike more complex nonlinear analyses, CSM uses simplified representations of both structural capacity and earthquake demand, making it a valuable tool for preliminary assessments and design checks, especially for those new to the field. It efficiently bridges the gap between the complexities of nonlinear analysis and the simplicity of linear methods. This makes it particularly suitable for preliminary design evaluations and quick checks on existing structures. Think of it as a first-pass filter, identifying potential weaknesses before committing to more intensive and costly analyses.

Its core principle involves comparing the structure's capacity, expressed as a capacity spectrum, with the expected earthquake demand, represented by a demand spectrum. The capacity spectrum illustrates the structure's resistance to seismic forces at different displacement levels. It's determined through simplified nonlinear analysis methods, often involving static pushover analysis, a topic covered in the preceding chapter. This analysis progressively applies lateral forces to the structural model until a predefined performance level is reached. Key here is understanding that the capacity spectrum is not a single value but a curve reflecting the structure's behavior across a range of displacements.

The demand spectrum, conversely, represents the earthquake's expected intensity at various structural periods. This is derived from ground motion records, scaled to a specific return period reflecting the desired level of seismic hazard. In essence, the demand spectrum summarizes the anticipated seismic forces the structure will experience during an earthquake of a specified magnitude. Therefore, accurately defining both the capacity and demand spectra is critical for obtaining a reliable assessment of the building's performance. Inaccurate estimations can lead to either an overestimation or underestimation of the building's capacity in comparison to the real earthquake.

Comparing the capacity and demand spectra is the heart of the CSM. If the capacity spectrum lies entirely above the demand spectrum, this suggests that the structure's strength exceeds the anticipated earthquake forces at all displacement levels, indicating acceptable performance. However, if the capacity spectrum falls below the demand spectrum at any point, this implies potential failure at that displacement level. This signifies areas needing further investigation and possibly retrofitting.

A critical aspect of the CSM is the selection of appropriate performance levels. These levels dictate the acceptable levels of damage under a given earthquake intensity. Typical performance levels include immediate occupancy (IO), life safety (LS), and collapse prevention (CP). IO signifies minimal damage, allowing immediate continued occupancy after the event. LS requires the structure to withstand the event without collapse, though damage may be extensive, requiring repairs. CP, the strictest level, aims to prevent total collapse, even under extreme seismic events. Choosing the appropriate performance level directly influences the design criteria and the interpretation of the CSM results. The selection of performance levels often depends on the intended use of the building, and the risk associated with failure of the structure.

Several factors can influence the accuracy and reliability of the CSM. The selection of the ground motion record significantly affects the demand spectrum, so using representative records is crucial. Likewise, the accuracy of the pushover analysis directly impacts the capacity spectrum. Careful modeling of the structure, including material properties and nonlinear behavior, is essential for obtaining a realistic capacity spectrum. Furthermore, the simplified nature of the CSM means it may not fully capture all the complexities of nonlinear structural behavior, meaning its results should always be considered with appropriate engineering judgment.

In conclusion, the Capacity Spectrum Method provides a valuable and efficient tool for seismic performance evaluation. Its simplified approach makes it readily accessible for beginners while still offering significant insights into a structure's behavior under seismic loading. However, remembering the limitations and understanding the underlying assumptions are essential to interpreting the results correctly. While offering a first-pass assessment, CSM should be used in conjunction with other methods, as highlighted in subsequent chapters, for a comprehensive seismic evaluation. The combination of CSM, pushover analysis, and other detailed analysis methods offers a robust framework for evaluating seismic safety and planning effective seismic retrofitting strategies. This holistic approach ensures that the design and retrofitting of structures maximize their resilience against seismic hazards, thereby contributing to enhanced building safety and community well-being. The practical implications are clear: improved life safety, reduced economic losses, and the creation of more resilient built environments. Understanding CSM is a crucial step towards this goal. Remember that while the CSM provides a valuable tool for assessing seismic performance, it is important to consider its limitations and use it in conjunction with other advanced analysis methods for a comprehensive evaluation. This chapter serves only as an introduction to this important method; a deeper understanding requires further study and hands-on application.

3.3 Pushover Analysis

Pushover analysis is a powerful tool for evaluating the seismic performance of structures, particularly in the context of seismic retrofitting. Unlike linear analyses that assume elastic behavior, pushover analysis considers the nonlinear behavior of materials and components, providing a more realistic representation of a building's response during an earthquake. It's a simplified approach, yet offers invaluable insights for decision-making in improving seismic resilience.

The process typically involves applying a gradually increasing lateral load to a structural model, simulating the effects of an earthquake. This load is applied incrementally, and at each step, the model's response is calculated, accounting for material nonlinearities like yielding of steel or cracking of concrete. The analysis continues until the structure reaches its ultimate capacity, meaning it has essentially failed in a certain way. This failure point provides crucial information about the structure's strength and ductility. The results are presented graphically as a pushover curve, showing the relationship between base shear and roof displacement.

One of the key benefits of pushover analysis is its ability to identify the weakest links within a structure. By observing which components fail first and the sequence of failure, engineers can target their retrofitting efforts effectively. For example, a pushover analysis might reveal that a specific column is prone to buckling before the rest of the structure fails. This information allows for targeted strengthening of that column, maximizing the benefits of the retrofit. This targeted approach is far more efficient than indiscriminately strengthening the whole structure.

However, pushover analysis does have limitations. Its accuracy depends heavily on the quality of the structural model, which should accurately represent the building's geometry, material properties, and connection details. Inaccuracies in the model can lead to misleading results, making careful model creation crucial. The method also simplifies the complex dynamic nature of earthquakes by applying a static load pattern. While it provides valuable information on the structure's strength and ductility, it doesn't capture the dynamic effects like resonance or higher mode effects that could be important for understanding the overall response.

Despite these limitations, pushover analysis remains a cornerstone of seismic design and retrofitting. It offers a practical and relatively inexpensive approach compared to more complex nonlinear dynamic analyses. This makes it particularly valuable for preliminary assessments and the initial phases of a retrofitting project, guiding subsequent stages such as detailed design. It's frequently used alongside other analysis methods, providing a comprehensive picture of the structure's seismic performance. For example, combining pushover results with capacity spectrum analysis allows engineers to determine the building's performance level against a range of seismic ground motions.

Many software packages are available for performing pushover analysis. These tools simplify the process, allowing engineers to input the structural model and obtain the results efficiently. Remember, however, that software is merely a tool; the interpretation of results requires significant expertise and judgment. The accuracy of the analysis hinges on the expertise used to create the model and to interpret the results. A skilled engineer is essential to ensure that the analysis accurately reflects the building's actual behavior and to draw meaningful conclusions for the retrofit strategy. Improper application can lead to ineffective and potentially dangerous retrofitting solutions.

Understanding the limitations and correctly interpreting the results is paramount. Don't treat it as a definitive answer, but rather as a valuable tool to inform design decisions. For instance, a pushover curve might show a brittle failure mode, indicating a need for ductile detailing in the retrofit design. Conversely, a ductile failure mode could suggest the focus should be on increasing the structure's overall strength. The analysis should therefore be considered within a larger context of overall structural assessment, and it shouldn't be interpreted in isolation. Using pushover analysis effectively requires understanding the principles of structural mechanics, nonlinear behavior of materials, and the limitations inherent in any simplified analysis technique. This informed use is fundamental to its success in improving seismic resilience. Therefore, always ensure you have a qualified structural engineer to undertake this work and interpret the complex results.

3.4 Performance-Based Design

Performance-based design moves beyond simply meeting minimum code requirements; it aims to achieve a specific performance objective under seismic loading. Instead of a prescriptive approach, it focuses on defining acceptable levels of damage and ensuring the structure can withstand anticipated seismic events without catastrophic collapse. This shift in philosophy requires a more sophisticated understanding of structural behavior and a detailed assessment of potential hazards.

Unlike traditional methods that primarily focus on preventing collapse, performance-based design allows for controlled damage. This means accepting a degree of non-structural damage while safeguarding the structural integrity and ensuring life safety. The design process considers the building's intended use, occupancy, and the potential consequences of failure. For instance, a hospital requires a higher level of performance than a warehouse. This nuanced approach necessitates a comprehensive evaluation of risk.

A key element of performance-based design is the use of advanced analytical techniques. Pushover analysis, for example, simulates the building's response to increasing lateral forces, providing insight into its capacity and potential failure mechanisms. Nonlinear dynamic analysis offers a more detailed representation of the structure's behavior during an earthquake, capturing the complex interaction between the structure and the ground motion. These analyses aren't merely theoretical exercises; they inform crucial decisions about design modifications and material selection.

Selecting appropriate materials is crucial in achieving desired performance levels. High-strength steel and ductile concrete, for example, are capable of absorbing significant energy during seismic events. Fiber-reinforced polymers (FRP) provide an effective means of strengthening existing structures, enhancing their capacity and ductility. The choice of material hinges on several factors, including cost-effectiveness, availability, and compatibility with existing construction. Careful consideration of these factors is essential for successful implementation.

The performance-based approach demands thorough documentation and verification. This includes detailed design calculations, analysis results, and construction specifications. Regular inspections during construction are vital to ensure compliance with the design intent. Post-construction evaluation, possibly including monitoring systems, provides valuable data for validating the design's performance and identifying areas for future improvement. The data gathered serves as feedback for iterative enhancements.

Successful implementation of performance-based design requires collaboration among engineers, architects, contractors, and building owners. Open communication and a shared understanding of the design goals are critical. A thorough understanding of the local seismic hazard and the building's specific vulnerability is paramount. By working together, a team can develop a design that balances cost, safety, and functionality, leading to a more resilient and safer building. This collaborative approach is not simply a recommendation, but a necessity for a successful project.

Finally, the adoption of performance-based design necessitates a change in mindset. It requires moving away from a purely prescriptive approach towards a more performance-oriented one. This means embracing uncertainty, considering potential failure modes, and focusing on minimizing the consequences of damage rather than solely preventing it. It's about designing for resilience, not just for compliance. This paradigm shift allows for innovation and improved safety standards in seismic design. The acceptance of calculated risk is a cornerstone of this evolved approach. It's about understanding and managing risks within acceptable performance levels.

3.5 Advanced Analytical Techniques

Nonlinear dynamic analysis goes beyond the limitations of simpler methods. Unlike linear analysis, which assumes a proportional relationship between load and displacement, nonlinear analysis accounts for material nonlinearity, geometric nonlinearity, and the complex interaction between structural components under severe seismic loading. This allows for a more accurate prediction of a structure's behavior, capturing crucial aspects like yielding, cracking, and damage progression, which are often critical in evaluating seismic performance. It's a powerful tool for understanding how a structure might respond to a strong earthquake, going beyond simple pass/fail criteria to offer a much richer understanding of its performance.

Finite element analysis (FEA) is a cornerstone of advanced seismic analysis. FEA discretizes a structure into numerous smaller elements, allowing for the detailed modeling of complex geometries and material properties. This level of detail is crucial for capturing the localized effects of seismic loading, including stress concentrations and potential failure points. Sophisticated FEA software packages allow engineers to simulate the dynamic response of structures to various earthquake ground motions, providing detailed information on displacements, stresses, and strains throughout the structure. Analyzing these results allows for the precise identification of weak points and vulnerabilities, informing targeted retrofitting strategies. The process demands significant computational power and expertise in model creation and interpretation.

Time-history analysis provides a realistic simulation of a structure's response to a recorded ground motion. Unlike simpler methods that use idealized ground motions, time-history analysis uses actual earthquake records, capturing the unique characteristics of specific seismic events. This detailed approach provides valuable insights into the structure's dynamic behavior under realistic loading conditions, accounting for the complex time-dependent aspects of earthquake ground shaking. The results are time-dependent, offering a chronological view of displacements, accelerations, and internal forces, allowing a deeper understanding of the structure's response during the earthquake's various stages. This level of detail is indispensable for performance-based design and critical for structures where safety is paramount.

Advanced numerical methods, such as the arc-length method and the Newton-Raphson method, are essential for solving the complex nonlinear equations that arise in advanced seismic analysis. These iterative solution techniques are crucial for accurately tracking the structure's response as it undergoes significant deformation and damage. The selection of an appropriate numerical method depends on factors such as the structure's complexity and the desired level of accuracy. Careful consideration of these methods and their convergence characteristics is vital for ensuring the accuracy and reliability of the analysis results. Understanding the limitations and strengths of each method empowers engineers to make informed decisions throughout the analysis process.

Probabilistic seismic demand analysis (PSDA) incorporates the inherent uncertainties associated with seismic hazards and structural response. Unlike deterministic methods that rely on single values for earthquake ground motions and structural properties, PSDA considers the variability of these parameters using probabilistic models. This approach yields a more comprehensive understanding of the potential range of structural responses, including the probability of exceeding specific performance levels. By quantifying the uncertainty associated with seismic events and structural behavior, PSDA provides a more realistic assessment of seismic risk, informing decision-making in design and retrofitting strategies. This probabilistic approach contributes to robust and reliable decision-making within the context of inherent uncertainty in seismic phenomena. The combination of these advanced analytical techniques provides a powerful toolkit for seismic design and retrofitting. By integrating nonlinear dynamic analysis, FEA, time-history analysis, sophisticated numerical methods, and PSDA, engineers can perform comprehensive assessments of structural performance under seismic loading. The insight gained from this rigorous analysis enables the development of effective retrofitting strategies that enhance the seismic resilience of existing structures and minimize potential damage and loss of life. This comprehensive approach is essential for ensuring public safety and protecting critical infrastructure in earthquake-prone regions. The choice of the most appropriate techniques depends greatly on the specific project requirements, the available resources, and the desired level of accuracy in the analysis.

4 Seismic Retrofitting Techniques

Strengthening masonry walls, a prevalent feature in older buildings, often requires a multifaceted approach. We can't simply slap on some new material; a thorough assessment of the existing wall's condition, including the mortar's strength and the presence of cracks, is paramount. This informs the selection of appropriate strengthening techniques. Common methods include the application of external reinforcement, like steel plates or fiber-reinforced polymers (FRP). These materials enhance the wall's tensile strength, a crucial factor in resisting earthquake-induced forces. Furthermore, we might inject epoxy resins into cracks to improve the mortar's bonding capacity, thereby increasing the wall's overall stability and load-bearing capabilities. In some cases, internal strengthening might be necessary, involving the installation of steel reinforcement within the wall's core. This requires careful planning to minimize disruption and ensure structural integrity. The choice of reinforcement method depends on factors such as the wall's thickness, material properties, and the level of seismic risk. Careful consideration of these factors, and expert consultation, is always needed.

Improving foundation stability is another critical aspect of seismic retrofitting, often overlooked. Shallow foundations, common in many older structures, are particularly vulnerable during earthquakes. Ground shaking can cause significant settlement or even foundation failure. Therefore, strengthening these foundations is key to improving a building's overall seismic performance. Several effective techniques exist to achieve this. One solution is to underpin the existing foundation with new, deeper footings or piles. This method transfers the building's weight to a more stable soil layer, reducing the risk of settlement. Alternatively, foundation strengthening can involve the use of ground improvement techniques, such as soil compaction or grouting, to increase the soil's bearing capacity. The specific technique chosen depends on the soil type, groundwater conditions, and the foundation's existing design. For example, where soil liquefaction is a concern, methods like dynamic compaction or vibro-compaction can be employed to densify the soil and reduce the risk. Remember that proper drainage is essential to prevent water accumulation around the foundation, which can lead to settlement and instability.

The effectiveness of any retrofitting technique relies heavily on proper material selection. Concrete, a fundamental construction material, demonstrates different responses under seismic loads based on its compressive and tensile strengths. High-strength concrete offers enhanced resistance to compression, but it may still be vulnerable to tensile stresses during seismic events. Therefore, the use of reinforcing steel is frequently necessary. The selection of the correct grade of concrete and steel must consider the anticipated seismic forces, the building's age and condition, and the available construction methods. Further materials such as FRP composites, offering high tensile strength-to-weight ratios, are increasingly employed. They have excellent durability and can be easily applied to existing structures, minimizing disruption to the building's occupancy. However, careful consideration is needed for factors such as their long-term performance and the possible degradation due to ultraviolet radiation or environmental exposure. Therefore, a thorough evaluation and understanding of each material's properties and limitations are necessary to achieve optimal results.

Detailed site investigations, including thorough geotechnical studies, play a critical role in the selection of an appropriate retrofitting approach. This includes determining the soil's bearing capacity, liquefaction potential, and groundwater conditions. These investigations not only inform the selection of the most effective foundation improvement techniques but also assist in evaluating the overall seismic vulnerability of the building. For instance, the presence of soft soils can significantly amplify ground motion, increasing the seismic demands on the structure. Similarly, the existence of underground utilities needs to be assessed to plan accordingly, avoiding interference during the retrofitting work. Therefore, a complete understanding of the site conditions is crucial for developing a cost-effective and effective seismic upgrade. This detailed assessment will guide the design process and help to ensure the longevity of the retrofitting solution. Remember, a well-informed design choice will minimize any risks.

Finally, the successful implementation of any seismic retrofitting strategy demands rigorous quality control throughout the entire process. This begins with meticulous planning and design, following through to the execution of the work itself, and is finally concluded with thorough post-retrofit inspections. This ensures that the retrofitted structure meets the required standards and performs as intended. Regular monitoring and maintenance should also be considered to ensure long-term performance and prevent future degradation of the structural integrity. The adherence to appropriate safety protocols during construction is also crucial. Protecting both construction workers and the public is of utmost importance. Regular inspections by qualified professionals allow for early identification of any problems and provide necessary guidance on potential remediation measures to ensure success.

4.1 Strengthening Masonry Walls

Masonry walls, often found in older buildings, are particularly vulnerable during seismic events. Their inherent weakness lies in the lack of ductility and the potential for brittle failure under lateral loads. Strengthening these walls is crucial for enhancing the seismic resilience of existing structures. This process involves various techniques, each tailored to the specific condition and material properties of the wall.

A common approach involves the use of external strengthening techniques. This could involve adding reinforced concrete jackets or steel bracing systems to the exterior faces of the walls. The reinforced concrete jackets effectively confine the masonry, increasing its shear strength and overall stability. Steel bracing, on the other hand, provides lateral support and prevents out-of-plane movement during seismic shaking. Careful design considerations are essential to ensure proper load transfer and avoid damaging the existing masonry during installation. This requires meticulous planning and precise execution to avoid unintended consequences.

Internal strengthening methods focus on reinforcing the masonry from within. This can be achieved through the injection of grout or epoxy resins into the existing mortar joints. These materials improve the bond between the bricks or stones, thus increasing the wall's tensile and shear strength. Inserting steel reinforcement elements, such as ties or bars, into the wall's core also enhances its ductility and capacity to withstand lateral forces. These internal interventions are less visible than external techniques, maintaining the original aesthetic appeal of the structure, making them especially attractive for historically significant buildings.

Another effective technique is the use of fiber-reinforced polymer (FRP) composites. FRP materials, known for their high strength-to-weight ratio, can be applied to the wall's surface in various forms, like sheets or wraps. These composites significantly improve the tensile strength of the masonry, preventing cracking and reducing the risk of collapse. Their light weight further minimizes additional load on the structure. Proper surface preparation and adhesion are critical for successful FRP application, warranting the use of appropriate primers and bonding agents to ensure long-term durability and effectiveness under seismic stress.

The selection of the most appropriate strengthening method depends on several factors. These factors include the wall's existing condition, the severity of the seismic hazard, the building's architectural significance, and, of course, the available budget. A comprehensive assessment by a qualified structural engineer is essential to determine the optimal solution. The engineer should carefully evaluate the wall's material properties, geometry, and load-bearing capacity to identify its weaknesses and tailor the retrofitting strategy accordingly. Cost-benefit analysis and potential disruption to building occupants should also be taken into account.

Finally, post-strengthening evaluation is crucial to ensure the effectiveness of the retrofitting measures. This can involve non-destructive testing, such as ultrasonic pulse velocity tests, to assess the condition of the masonry and the integrity of the strengthening materials. Detailed load tests, either physical or numerical, may be employed to validate the wall's improved load-carrying capacity and its seismic performance. Documentation of the entire process, including design calculations, material specifications, and construction details, is necessary for future reference and maintenance. These comprehensive steps guarantee the longevity and efficiency of the seismic retrofitting efforts, ensuring the building's safety and structural integrity for years to come. Regular inspections are highly recommended. Ignoring this step could compromise the investments made in this important work. This ensures that the structure remains resilient to future seismic events. Safety is paramount, after all.

4.2 Improving Foundation Stability

Shallow foundations, common in many older buildings, are particularly vulnerable during seismic events. Their susceptibility to uplift, sliding, and settlement can compromise the entire structure's stability. Improving their resilience requires a multifaceted approach, carefully considering soil conditions, foundation type, and the building's overall design. Ignoring these factors can lead to catastrophic failures, even with seemingly minor ground shaking. Therefore, a thorough site investigation is paramount.

Soil testing is crucial to determine the soil's bearing capacity, shear strength, and liquefaction potential. These parameters dictate the appropriate foundation improvement strategy. Liquefaction, where saturated sandy soils lose strength under seismic shaking, is a particularly insidious threat. Identifying areas prone to liquefaction allows for preventative measures, such as soil densification or ground improvement techniques. A detailed geotechnical report should inform all subsequent decisions.

Foundation strengthening often involves increasing the foundation's overall bearing capacity. This can be achieved through various methods, depending on the existing foundation and soil conditions. Underpinning, a process of extending the existing foundation deeper into more stable soil layers, is a common technique. This provides increased resistance to uplift and settlement. Another approach involves adding additional foundation elements, like pile foundations or drilled shafts, to augment the load-carrying capacity. The selection of method will depend on several factors.

Improving the foundation's resistance to sliding involves increasing the frictional resistance between the foundation and the soil. This can be achieved by increasing the foundation's weight or by improving the soil's shear strength. Adding counterforts or buttresses to the foundation can significantly enhance its lateral resistance. Alternatively, ground improvement techniques, such as grouting or soil stabilization, can improve the soil's shear strength and reduce the potential for sliding. These approaches complement each other.

For foundations exhibiting significant settlement, raising the foundation might be necessary. This involves carefully lifting the structure and installing new foundation elements. This is a complex and costly procedure that requires expertise and careful planning to avoid further damage. This method is generally used as a last resort, when other, less invasive solutions prove insufficient. Careful monitoring is crucial throughout.

The selection of materials for foundation strengthening is crucial. High-strength concrete and reinforced steel are commonly used, ensuring adequate durability and load-bearing capacity. Geosynthetics, such as geogrids and geotextiles, can also be utilized to enhance soil stability and prevent settlement. The choice of materials should reflect both technical requirements and economic considerations. Material properties must be carefully considered.

Beyond material selection, proper detailing and construction practices are paramount. The quality of workmanship directly impacts the foundation's ultimate performance. Strict adherence to design specifications is essential to ensure that the strengthening work effectively enhances the foundation's resistance to seismic loads. Negligence in this phase can negate all other efforts. Regular inspections are vital.

Finally, regular inspection and maintenance are crucial to ensuring the long-term stability of the retrofitted foundation. Regular monitoring, ideally through instrumentations such as inclinometers and settlement gauges, can detect any subtle changes in foundation behavior. Early detection of potential problems allows for timely intervention, preventing catastrophic failures. Proactive maintenance is essential for preserving the seismic resilience improvements. This ensures continued protection.

5 Material Selection & Properties

The performance of any retrofitted structure hinges critically on the properties of the materials used. Concrete, a ubiquitous material in construction, exhibits complex behavior under seismic loading. Its compressive strength is generally high, but its tensile strength is significantly lower, making it vulnerable to cracking. Understanding this inherent weakness is paramount when selecting concrete for seismic retrofitting projects. We must carefully consider the concrete's mix design, ensuring sufficient strength and ductility to withstand anticipated seismic forces. Careful consideration of factors like aggregate type, cement content, and water-cement ratio becomes crucial in achieving the desired performance characteristics. Poorly mixed concrete can lead to premature failure, undermining the entire retrofitting effort. Therefore, rigorous quality control procedures are essential throughout the entire process, from material sourcing to final placement. Independent testing of concrete samples is highly recommended to verify that the concrete meets the specified strength and durability requirements.

Steel, another mainstay in construction, offers high tensile strength and ductility, making it ideal for reinforcing concrete structures and providing crucial structural support during earthquakes. However, steel's susceptibility to buckling under compression must be carefully managed through appropriate design and detailing. Proper connection details are vital to ensure that the steel reinforcement effectively transfers forces to the surrounding concrete, preventing premature failure. Furthermore, the detailing must account for potential corrosion, a factor that can significantly reduce steel's long-term strength. Corrosion protection measures such as galvanizing or using stainless steel are often necessary, particularly in environments with high humidity or exposure to de-icing salts. This careful selection and implementation are essential to ensure that the steel reinforcement enhances seismic resistance effectively and contributes to the longevity of the reinforced structure. The cost-benefit analysis of using different grades of steel should be considered.

Fiber-Reinforced Polymers (FRPs) have emerged as a valuable addition to the retrofitting arsenal. These composite materials combine high strength-to-weight ratios with excellent durability. FRPs are particularly well-suited for strengthening existing structures with minimal disruption. They can be easily bonded to existing concrete or masonry, providing additional tensile strength and improving the structure's overall ductility. However, FRPs are susceptible to ultraviolet (UV) degradation, requiring protective coatings or careful placement to maintain their long-term performance. Moreover, proper surface preparation of the substrate is critical for ensuring a strong bond between the FRP and the existing structure. Failure to achieve adequate bonding can lead to premature debonding and compromise the effectiveness of the retrofit. Therefore, meticulous application procedures and careful selection of appropriate adhesive systems are paramount when using FRPs in seismic retrofitting. Understanding the different types of FRP (carbon, aramid, glass) and their respective properties is crucial for making informed material selection decisions.

Beyond concrete, steel, and FRPs, other advanced materials are finding increasing use in seismic retrofitting. These include shape memory alloys (SMAs), which can recover their original shape after deformation, and self-healing materials, which can repair microscopic cracks autonomously. While these materials present exciting possibilities for improving the resilience of structures, their application is currently limited by their cost and the complexity of their implementation. Further research and development are needed to overcome these challenges and make these advanced materials more readily available for widespread adoption in seismic retrofitting. However, the potential benefits, such as enhanced energy dissipation and increased structural longevity, warrant continued exploration and refinement. This exploration into newer materials should focus on

their sustainability and their life-cycle environmental impact.
The selection of materials for seismic retrofitting requires a holistic approach, considering not only the material's mechanical properties but also its durability, cost-effectiveness, and ease of installation. The structural engineer needs to conduct a thorough assessment of the existing structure, considering its age, condition, and the specific seismic hazards it faces. This assessment informs the choice of materials, ensuring that they are appropriate for the specific application and will provide the necessary level of seismic protection. Furthermore, it is crucial to consider the long-term maintenance implications of the chosen materials and their potential impact on the building's overall lifespan. Careful consideration of material properties, application methods, and long-term performance is essential to ensure the success of the retrofitting project and the safety of the building's occupants. Remember, selecting the right material is not just about strength; it's about achieving a balance between strength, durability, cost, and ease of implementation to create a truly resilient structure. Only by understanding these factors fully can engineers successfully enhance building safety in seismically active regions.

5.1 Concrete & Steel Behavior

Concrete, a ubiquitous material in construction, exhibits complex behavior under seismic loading. Its compressive strength is significantly higher than its tensile strength, a crucial factor in seismic design. Ductility, the ability to deform plastically before failure, is limited in plain concrete, making it prone to brittle fracture under strong earthquake shaking. Reinforcement with steel significantly enhances this ductility, allowing for energy dissipation and increased resilience. However, the bond between concrete and steel is critical and its degradation, even slight, drastically weakens the entire structural element. Understanding this interaction, and potential failures at the interface due to corrosion or cracking, is paramount to designing effective and long-lasting seismic retrofits. Proper detailing, including sufficient embedment lengths and confinement, is therefore vital for ensuring reliable performance.

Steel, on the other hand, possesses high tensile strength and ductility, making it an ideal material for reinforcing concrete structures. Its yield strength, the point at which it begins to deform plastically, is a key parameter in seismic analysis. During an earthquake, steel's ductility allows it to absorb significant energy, preventing catastrophic failure of the concrete structure. However, steel's susceptibility to buckling under compression necessitates careful design considerations, particularly in columns. Buckling, a sudden failure mode, can be mitigated by using high-strength steel or incorporating lateral bracing. The inherent brittleness of welds compared to the parent steel must also be factored in during seismic design and evaluation. Therefore, detailed welding procedures and quality control are non-negotiable.

The interaction between concrete and steel under seismic loads isn't simply additive. It's a dynamic interplay where their respective properties are amplified or mitigated based on their arrangement and the level of confinement provided. For instance, proper confinement of concrete using transverse reinforcement significantly improves its ability to withstand compressive stresses. This confinement, often achieved with closely spaced stirrups, hinders the expansion and cracking of the concrete, preventing premature failure. Similarly, the distribution of reinforcement influences the overall behavior. A poorly detailed reinforcement layout can lead to localized stress concentrations, resulting in early cracking and eventual failure. Conversely, a well-distributed reinforcement pattern can distribute stresses more evenly, increasing the structure's overall capacity and resilience. Advanced techniques, such as using high-performance concrete or fiber-reinforced concrete, further enhance the performance.

Understanding the material properties is only half the battle. The actual behavior under dynamic loading, as experienced during an earthquake, is significantly more complex. Factors such as concrete creep (time-dependent deformation), shrinkage (volume reduction due to drying), and the deterioration of steel due to corrosion all affect the long-term performance of the retrofitted structure. Therefore, it's crucial to utilize advanced analysis techniques, as discussed in previous chapters, that account for these time-dependent effects.

These advanced analyses might involve sophisticated simulations or experimental evaluations, allowing for a better prediction of the building's response to earthquake forces. Accurate prediction of this behaviour is crucial in ensuring that the strengthened structure effectively withstands the intended loads for its designated lifespan, particularly where the seismic activity is significant.

Seismic retrofitting necessitates a thorough understanding of the concrete and steel behavior under both static and dynamic conditions. It's not simply about strengthening the existing structure; it's about strategically enhancing its capacity to dissipate energy and resist seismic forces. This requires a detailed assessment of the current structural condition, identifying weak points, and selecting appropriate retrofitting techniques to address those deficiencies. Consider, for example, the use of externally bonded fiber reinforced polymers (FRP) to enhance the tensile strength of deficient concrete members. This is one of many approaches. The choice of techniques and materials must be based on a thorough understanding of the expected seismic demands and a detailed assessment of both material properties and structural performance. It's a meticulous process that requires both theoretical knowledge and practical experience in this engineering field. The end result should be a structure significantly enhanced in its seismic resilience, improving safety for all occupants. Selecting the right materials and understanding the interaction between the components of a structure leads to a safer and more resilient built environment.

5.2 FRP Composites in Retrofitting

Fiber Reinforced Polymers (FRP) offer a compelling solution for seismic retrofitting, particularly in situations where minimizing disruption and weight is crucial. Their high strength-to-weight ratio allows for substantial strengthening without adding significant mass to the structure, a key advantage in older buildings where excessive weight could exacerbate existing problems. Furthermore, FRP materials are readily adaptable to various shapes and structural components, enabling their integration into complex existing designs. This adaptability makes them incredibly versatile compared to traditional strengthening methods which often require extensive demolition and reconstruction.

The application of FRP composites is, however, not without its challenges. Proper surface preparation is paramount for ensuring effective bonding between the FRP material and the existing substrate. Contamination, moisture, and inadequate cleaning can significantly compromise the bond strength, thereby undermining the effectiveness of the retrofit. Furthermore, long-term durability is a critical consideration. The susceptibility of FRP to ultraviolet (UV) degradation and potential chemical attack necessitates the use of appropriate protective coatings. Thorough planning and careful selection of materials and coatings are essential to ensure the long-term performance of the retrofit. Neglecting this crucial aspect could lead to premature failure, defeating the purpose of the reinforcement.

Successful FRP retrofitting projects often rely on a meticulous understanding of the structural behavior under seismic loads. Advanced analytical techniques, such as finite element analysis (FEA), are commonly used to predict the performance of the retrofitted structure. FEA allows engineers to simulate earthquake loading and evaluate the stress distribution within the strengthened components. This detailed analysis helps in optimizing the design of the FRP reinforcement, ensuring that the material is placed strategically to effectively counter the anticipated seismic forces. It also aids in predicting potential failure modes and adjusting the design accordingly to enhance overall safety and resilience. Without rigorous testing and analysis, the use of this strong material can result in costly errors.

Beyond the technical aspects, successful implementation necessitates careful planning and project management. Coordination with contractors, ensuring quality control during installation, and adhering to strict safety protocols are all crucial aspects. The unique properties of FRP, such as their relatively high sensitivity to impact damage during installation, require highly skilled and experienced labor. Training and certification programs for contractors are thus essential to ensure appropriate installation practices and enhance the reliability of the retrofitted structure. Failure to do so could result in unexpected failure.

Furthermore, the economic viability of FRP retrofitting must be evaluated carefully. While FRP materials can be cost-effective in certain situations, the overall cost-benefit analysis needs to account for various factors including material costs, labor costs, design and analysis expenses, and potential disruption to building use. A comprehensive cost analysis, coupled with an assessment of the building's expected lifespan and potential risks, will provide crucial information for decision-making. A comprehensive approach minimizes the chance of unintended consequences in the process.

The selection of the appropriate FRP type is equally important. Different FRP materials, such as carbon fiber reinforced polymers (CFRP) and glass fiber reinforced polymers (GFRP), exhibit different mechanical properties and are suitable for various applications. CFRP generally possesses higher strength and stiffness, making it ideal for high-strength reinforcement, while GFRP offers a more cost-effective solution. The choice of FRP material must align with the specific requirements of the structure and the nature of the anticipated seismic loads. A careful evaluation of the building's structural characteristics ensures the right choice.

Finally, post-retrofit monitoring is indispensable for validating the effectiveness of the implemented solution. Regular inspections and potentially advanced monitoring systems using sensors to track strain and displacement can provide valuable data on the long-term performance of the retrofitted structure. This data can be utilized to refine future design approaches and help to better understand the long-term behavior of these materials in real-world conditions. Such observation over time enhances the understanding of FRP performance, leading to continuous improvement in future projects. This systematic feedback loop ensures continued development in the use of FRP composites in seismic retrofitting and provides an assurance to concerned parties about the sustainability and efficiency of FRP.

5.3 Other Advanced Materials

Self-healing concrete represents a significant advancement. Imagine a material that can automatically repair minor cracks, extending its lifespan and reducing the need for frequent maintenance. This is the promise of self-healing concrete, achieved through the incorporation of bacteria or other self-healing agents that react to cracks, producing calcium carbonate to fill the gaps. Its use in seismic retrofitting offers a compelling pathway to long-term structural integrity. This innovative approach minimizes the need for extensive repairs post-earthquake, thus saving time and money while enhancing the structure's resilience over decades.

Shape memory alloys (SMAs) are another game-changer. These metallic alloys exhibit a remarkable ability to "remember" their original shape, even after being deformed. When heated, they return to their pre-deformed state. In seismic retrofitting, strategically placed SMAs can act as dampers, absorbing seismic energy and reducing the impact of ground motion on the structure. While their implementation might involve complex integration, the potential for enhanced seismic resilience is undeniable, especially in areas where high-frequency vibrations are a concern. Research indicates that optimal placement and design of SMA components contribute significantly to their effectiveness. Their relatively high cost is a factor to consider but is often balanced by the long-term cost savings associated with reduced maintenance and damage.

Carbon nanotubes (CNTs) possess exceptional strength and flexibility, making them ideal candidates for reinforcing existing structures. Their incorporation into concrete or other composite materials can significantly enhance the overall tensile strength and ductility of the retrofitted structure. While the production of CNTs remains expensive, ongoing research is exploring more cost-effective manufacturing techniques. The potential for CNT-enhanced materials to improve the performance of aging structures during seismic events is significant. Preliminary studies demonstrate an increase in both tensile strength and flexural performance over traditional reinforcement approaches. However, practical implementation needs careful consideration of the dispersion and integration techniques to guarantee optimal enhancement.

Finally, let's consider the use of bio-based materials. This is a relatively new frontier in seismic retrofitting, but it offers significant potential environmental benefits. Bio-based materials, derived from renewable sources such as bamboo or hemp fibers, possess inherent strength and flexibility. When correctly treated and integrated, they offer a sustainable alternative to traditional reinforcing materials. This aligns with growing concerns about the environmental impact of conventional construction and offers a path towards eco-friendly seismic retrofits. Research into their long-term durability and performance under seismic loads is ongoing, but the initial results appear promising. The reduced carbon footprint and potential for local sourcing present compelling arguments for their broader consideration in the field. While bio-based materials are still a newer field and may require further research to fully realize their potential for large-scale projects, their sustainable nature and potential performance benefits are noteworthy. Their application, especially in smaller-scale projects or in regions with abundant local bio-resources, presents an intriguing and environmentally friendly approach.

6 Case Studies & Examples

A significant retrofit project involved a three-story reinforced concrete school building constructed in the 1960s. Located in a seismically active zone, the building showed signs of significant deterioration, including cracking in columns and beams, and inadequate shear capacity. The retrofit strategy focused on strengthening the columns using externally bonded fiber-reinforced polymer (FRP) sheets, a cost-effective and minimally disruptive solution. This strengthening increased the building's ductility and significantly improved its seismic performance, as verified through detailed nonlinear finite element analysis. Post-retrofit evaluations showed a substantial increase in the building's lateral load capacity.

Another compelling case study centers on a historic masonry church. Its age and construction methods left it extremely vulnerable to earthquake damage. The retrofit involved a carefully planned strategy combining internal strengthening with external buttressing. Internal strengthening used steel bracing to reinforce the existing load-bearing walls, preventing collapse. External buttressing, using reinforced concrete elements, provided additional support and increased the building's lateral stiffness. This approach balanced preservation of the historical character with the need for improved seismic performance, demonstrating the potential for blending historical preservation with modern engineering practices.

Consider the seismic upgrade of a multi-story steel-framed office building. This project highlighted the importance of addressing both structural and non-structural components. While the steel frame itself possessed some inherent ductility, non-structural elements like infill walls and cladding presented a significant risk of damage and potential collapse. The retrofit incorporated several strategies. The infill walls were strengthened using shear walls and bracing systems, preventing collapse and reducing the risk of injury. The cladding was also redesigned for improved resilience to seismic forces. This comprehensive approach underscored that effective seismic retrofitting needs a holistic perspective that considers all elements of the structure.

A different challenge was presented by a low-rise residential building constructed using unreinforced masonry. This structure was particularly vulnerable to collapse during an earthquake. The retrofit strategy opted for a combination of techniques, beginning with the introduction of shear walls to increase the building's lateral stiffness. Moreover, the foundation was improved using underpinning techniques to enhance its ability to resist seismic forces. This method improved the building's base strength, reducing the chances of foundation failure. This example underlines that even seemingly simple structures benefit greatly from tailored retrofitting techniques.

Finally, let's examine the retrofitting of a long-span concrete bridge. This project highlighted the specific challenges of retrofitting infrastructure. The bridge's aging concrete components were showing signs of deterioration and inadequate seismic capacity. A key element of the upgrade involved strengthening the bridge's piers using jacketing techniques, reinforcing the existing concrete with high-strength concrete and steel. Expansion joints were also upgraded to ensure continuity during seismic events. This case demonstrates the complexity of seismic retrofitting for infrastructure, often requiring the use of sophisticated engineering approaches and materials.

Each of these examples, diverse in structure and material, underscore the crucial need for a tailored, site-specific approach to seismic retrofitting. Successful projects require a thorough understanding of the building's structural characteristics, the local seismic hazard, and the selection of the most appropriate and cost-effective techniques.

Careful planning and execution are paramount to ensuring the safety and longevity of buildings in earthquake-prone regions. The importance of collaboration between engineers, architects, contractors and stakeholders cannot be overstated, ensuring that projects are executed efficiently and effectively, with due consideration for historical significance and potential disruptions.

6.1 Retrofit Project 1

This project involved a three-story unreinforced masonry (URM) building constructed in the 1920s, located in a zone of moderate seismic activity. The building, used as a commercial space, showed significant signs of age and deterioration, including cracked walls, uneven settlements, and a damaged foundation. Initial assessments revealed a high vulnerability to seismic forces, posing a considerable risk to occupants and the surrounding community. Our primary goal was to enhance the building's seismic performance while maintaining its historical character.

The initial steps involved a thorough structural evaluation. This included detailed inspections of the masonry walls, foundation, and roof structure. Non-destructive testing methods, such as ground-penetrating radar and ultrasonic pulse velocity tests, were employed to assess the condition of the materials and identify hidden damage. The data collected was used to create a detailed 3D model of the building, incorporating the identified defects and material properties. Advanced finite element analysis was then performed to simulate the building's response under various seismic scenarios. This allowed us to pinpoint the building's weakest points and estimate its potential damage under earthquake loading.

Based on the analysis results, we developed a comprehensive retrofitting plan. Our strategy centered on strengthening the URM walls through the installation of external steel bracing systems and the addition of reinforced concrete shear walls at critical locations. We also implemented foundation improvements, including underpinning and soil stabilization, to address the uneven settlement. The selected materials were chosen for their compatibility with the existing structure, strength, and durability. We prioritized materials that would minimize disruption to the building's aesthetic appeal and maintain the historical features.

The execution of the retrofitting plan was carefully managed. We began with the foundation improvements, underpinning the existing footings and stabilizing the soil to prevent further settlement. This crucial step ensured a stable base for the subsequent strengthening work. The installation of the steel bracing systems required meticulous precision, ensuring proper alignment and connection to the existing masonry. The team carefully considered the installation techniques to minimize damage to the original masonry and the impact on ongoing building operations. The construction schedule was planned to minimize disruption to the businesses operating within the building. Regular quality control checks ensured the work met the required standards.

Significant challenges arose during the project. The age and condition of the building presented unforeseen difficulties. Hidden structural weaknesses were revealed during demolition, requiring quick adaptations to the original plan. Coordination with the building's occupants to minimize disruptions proved crucial, and we implemented a phased approach to accommodate their needs while still maintaining a strict schedule. Open communication and thorough documentation were essential to ensure the project's successful completion. The close collaboration between the engineers, contractors, and building occupants was critical for navigating these unexpected issues. This collaborative approach ensured that everyone remained informed throughout the process.

The successful completion of this project demonstrated the effectiveness of carefully planned and executed seismic retrofitting. The improved seismic performance of the building not only protected the occupants but also added to the building's longevity. Post-retrofit assessments confirmed a significant increase in the building's lateral load capacity and overall seismic resilience. The building's vulnerability to seismic damage was substantially reduced, offering peace of mind to its occupants. This project highlighted the importance of combining detailed analysis with a pragmatic approach to retrofitting, resulting in a safe and resilient structure. The preservation of the building's historical character also showed that seismic retrofitting can be achieved without compromising the architectural integrity of significant structures. The project's success underscores the necessity for ongoing research and innovation in seismic retrofitting techniques to ensure the safety of our built environment. This approach minimizes risk and fosters a deeper understanding of preserving existing structures.

6.2 Retrofit Project 2

This project involved a mid-rise reinforced concrete office building constructed in the 1970s, exhibiting significant deficiencies in its seismic performance. Initial assessments revealed inadequate shear capacity in the columns, insufficient ductility in the beam-column joints, and a lack of proper detailing at critical connections. The building's location in a high seismic zone further exacerbated these concerns. Our team's primary objective was to enhance the building's ability to withstand seismic events while minimizing disruption to its ongoing operations.

The chosen retrofit strategy focused on strengthening the existing structural elements and improving their energy dissipation capabilities. We opted for a combination of techniques, carefully balancing cost-effectiveness with the required level of seismic performance enhancement. A key element of the strategy involved the installation of external steel bracing systems at strategic locations throughout the building's frame. This approach provided significant additional shear capacity and improved lateral stiffness, crucial for resisting seismic forces.

Careful consideration was given to the selection of materials. High-strength steel was employed for the bracing, ensuring optimal performance under extreme loads. The connection details were meticulously designed to guarantee the proper transfer of forces between the new bracing and the existing structure, avoiding potential failure points. Extensive testing and analysis were conducted to validate the design and to confirm that the retrofitted structure would meet the desired seismic performance targets. This included sophisticated nonlinear finite element analysis to assess the building's response under various seismic scenarios.

The retrofitting process itself was meticulously planned and executed to minimize disruption to the building's occupants and functionality. Work was carefully phased to ensure continuous building access and to maintain the comfort and safety of the building's tenants. Regular inspections and quality control checks were carried out throughout the construction process to guarantee that all work conformed to the approved design and specifications. We successfully completed the entire retrofit within the allocated budget and timeframe, minimizing inconvenience to building occupants.

Post-retrofit assessment demonstrated a significant improvement in the building's seismic performance. The building's capacity to withstand a major earthquake was markedly enhanced, significantly reducing the risk of collapse and mitigating potential losses. The improved seismic resilience provided peace of mind for both the building's owners and occupants, ensuring a safer environment for years to come. The project stands as a prime example of effective and practical seismic retrofitting, highlighting the importance of carefully planned interventions to safeguard existing infrastructure against seismic hazards. Our detailed analysis and meticulous execution were critical to the success of the project, demonstrating a commitment to safety and excellence. The comprehensive approach to the project highlights the importance of considering all aspects, from initial assessment and design, through material selection and construction, to final verification and evaluation. This case study serves as a robust illustration of best practices in seismic retrofitting, suitable for educational and professional contexts.

Furthermore, the successful completion of the project involved the skillful coordination of various stakeholders, including engineers, contractors, and building management. Effective communication and collaboration were paramount in ensuring the timely and efficient execution of the work. Regular meetings and progress reports were conducted to keep all parties informed of the project's status, which promoted collaboration and allowed for any potential challenges to be addressed proactively. These aspects underscore the importance of teamwork in ensuring the successful implementation of complex retrofitting projects. The lessons learned from this project can provide valuable insights for future seismic retrofitting endeavors, emphasizing the significance of integrated project management and thorough coordination.

The project also generated valuable data for future research and improved understanding of seismic retrofitting techniques. Data gathered during the assessment, design, and construction phases offers insights into the effectiveness of the selected materials and techniques. This information can be utilized in developing improved design guidelines and standards for future projects. The detailed post-retrofit assessment provided valuable information on the building's response to simulated seismic events, contributing to a more comprehensive understanding of seismic behavior in retrofitted structures. This rigorous approach to data collection and analysis is essential for advancing the field of seismic retrofitting and enhancing the safety of structures worldwide. The project's comprehensive documentation, including detailed drawings, specifications, and analysis reports, constitutes a valuable resource for future reference and research.

In conclusion, this project showcased the effective application of advanced seismic retrofitting techniques on an existing structure. The integrated approach to design, construction, and evaluation led to significant improvements in the building's seismic performance. The lessons learned from this experience offer valuable insights for professionals involved in seismic risk mitigation and strengthening existing buildings to better withstand future earthquakes. The comprehensive documentation and analysis resulting from this project provide a valuable case study for future research and practice. The detailed reporting and post-retrofit assessment provide valuable lessons, which can inform future endeavors. This project highlights the importance of a thorough and comprehensive approach to seismic retrofitting, leading to safer and more resilient buildings.

6.3 Retrofit Project 3

This project involved a mid-rise reinforced concrete office building constructed in the 1970s, exhibiting significant deficiencies in its seismic performance. The building's primary structural system consisted of a moment-resisting frame, but lacked sufficient ductility and shear capacity, particularly in the lower stories. Initial assessments using pushover analysis revealed a high probability of collapse under moderate to severe seismic events. We knew we had to act, and act decisively.

Our primary concern was the inadequate shear capacity of the columns, which were significantly under-designed compared to current seismic codes. The brittle nature of the concrete, compounded by inadequate detailing of the longitudinal and transverse reinforcement, posed a significant threat. We proposed a comprehensive retrofit strategy focusing on strengthening the columns and improving the overall ductility of the frame. This involved a multi-faceted approach, a blend of traditional and advanced techniques to assure success.

The selected solution involved the use of externally bonded fiber-reinforced polymer (FRP) jackets on the columns. The FRP composite material was chosen due to its high tensile strength, ease of application, and minimal disruption to the building's occupancy during the construction phase. Careful consideration was given to the bond strength between the FRP and the concrete, ensuring adequate transfer of forces. Extensive testing was conducted to determine the optimal FRP configuration and ensure compatibility with the existing concrete. This wasn't just about sticking on some material, it was about the scientific application.

In addition to the column strengthening, we also implemented improvements to the beam-column joints. These joints were another critical weakness in the original design, presenting vulnerable points in the structural system. We employed steel jacketing to improve the confinement of the concrete and enhance the joint's capacity to withstand shear forces. The steel jackets were carefully designed and installed to avoid stress concentrations and ensure proper integration with the existing structural elements. Meticulous detail was key to success. We couldn't afford to get this wrong.

Another important aspect of the retrofit involved the foundation. The original foundation, while structurally sound under typical loads, lacked the necessary ductility and capacity to withstand the increased seismic forces after the column strengthening. We employed additional ground improvement techniques to increase the soil's bearing capacity and reduce settlement potential. This involved installing soil nails to reinforce the underlying strata. This ensured the foundation could handle the added loads without compromising the building's stability. This was not a small undertaking. The entire retrofit process was meticulously documented, with regular inspections and testing conducted throughout. We utilized advanced monitoring technologies to track the building's response to dynamic loads. This provided valuable data for verifying the effectiveness of the retrofit and adjusting our strategy as needed. We employed non-destructive testing methods to assess the condition of the concrete before, during and after the strengthening works. This allowed us to fine-tune our approach and adjust our approach in real time as the process unfolded. It was important that all stages were tracked.

Post-retrofit analysis, including nonlinear dynamic analysis and pushover analyses, showed a significant improvement in the building's seismic performance. The improved ductility and shear capacity, thanks to the FRP jacketing and steel joint enhancements, resulted in a substantial increase in the building's collapse capacity. The enhanced foundation stability further ensured the integrity of the overall structural system. The building went from a likely collapse to performing exceptionally well under seismic load conditions. The success was measured not just by the enhanced structural performance, but also by the efficiency of the retrofit process, minimizing disruption to the building's occupants. It was a testament to meticulous planning and precision execution. It was truly a successful culmination of hard work and expert knowledge. The final report comprehensively details the methodology, results and lessons learned from this retrofit project, providing a valuable resource for engineers involved in similar projects. The success of this project reinforces the importance of careful planning and precision in seismic retrofitting. The focus on a multi-faceted approach, utilizing both traditional and advanced techniques, proved to be crucial in achieving the desired level of seismic resilience.

6.4 Retrofit Project 4

This project involved a mid-rise, unreinforced masonry (URM) building constructed in the 1920s, located in a zone of moderate seismicity. Its primary structural deficiency was the lack of ductile detailing and the inherent brittleness of the URM walls. Initial assessments revealed significant vulnerability to even moderate seismic shaking, posing a substantial risk to occupants. The building's age and architectural significance presented unique challenges, demanding a retrofit strategy that balanced seismic performance enhancement with preservation of its historical character.

The chosen approach prioritized strengthening the building's lateral load-resisting system. We employed a combination of techniques including the installation of externally bonded fiber-reinforced polymer (FRP) strips to enhance the shear capacity of the masonry walls. Careful consideration was given to the aesthetic impact of the FRP, opting for a color and texture that minimized visual disruption while ensuring robust performance. In addition to FRP strengthening, we implemented a new system of steel bracing within the interior walls to create a more unified structural system, distributing seismic forces more effectively. This minimized stress concentrations on the weaker masonry elements.

Detailed finite element analysis (FEA) was conducted to model the building's behavior under various seismic loading scenarios, both before and after the retrofit. The FEA results provided critical insights into the effectiveness of the chosen strengthening techniques, allowing us to fine-tune the design for optimal performance. This iterative design process ensured that the retrofit met stringent safety requirements without compromising the building's historical integrity. The analysis meticulously accounted for material nonlinearity, considering the brittle behavior of the masonry and the ductile response of the FRP and steel components. This accuracy was crucial for evaluating the overall structural capacity and predicting the building's response during a seismic event.

A key aspect of this project was the close collaboration with historical preservation experts. Their input ensured that the retrofitting interventions remained sensitive to the building's architectural features. The use of compatible materials and careful detailing were paramount in maintaining the building's original aesthetic appeal, demonstrating that seismic upgrading doesn't necessarily mean sacrificing historical charm. Regular consultations with the preservation team were essential throughout the project lifecycle, from the initial assessment to final inspection. Their detailed knowledge of the building's historical significance ensured the integrity of the structure was fully protected and preserved.

Construction procedures were meticulously planned to minimize disruption to the building's surroundings. Phasing the work allowed the building to remain partially operational throughout the retrofit, reducing economic impact on the tenants. The construction phase was managed with a focus on worker safety and compliance with building codes and industry best practices. This multi-faceted approach, emphasizing both structural resilience and historical preservation, has resulted in a retrofit project that serves as a successful case study in marrying engineering innovation with respect for cultural heritage. The post-retrofit structural analysis showed a significant improvement in the building's seismic performance, exceeding the minimum code requirements by a considerable margin. The comprehensive structural health monitoring continues to provide ongoing data, allowing for further refinements to future seismic retrofitting projects and enhancing our understanding of the long-term effectiveness of the employed techniques.

6.5 Retrofit Project 5

This project involved a mid-rise reinforced concrete office building constructed in the 1970s, exhibiting significant deficiencies in its seismic performance. Initial assessments revealed inadequate shear strength in the columns, a weakness exacerbated by the presence of brittle infill masonry walls. The building's foundation, although seemingly adequate, lacked sufficient ductility to withstand anticipated ground motions in the event of a major earthquake. These factors presented a significant risk to the building's occupants and structural integrity.

The retrofit strategy focused on a multi-pronged approach to address these shortcomings. We began by carefully removing the brittle infill masonry walls, replacing them with flexible, ductile framing systems designed to absorb seismic energy. This minimized the potential for catastrophic wall collapse during shaking, a common failure mode in older buildings. This process required meticulous planning and execution to ensure the structural integrity of the building was maintained throughout the demolition and reconstruction phases.

Strengthening of the existing reinforced concrete columns was achieved using externally bonded fiber-reinforced polymer (FRP) sheets. This technique, chosen for its cost-effectiveness and minimal disruption to building operations, significantly enhanced the columns' shear capacity and ductility. The application of the FRP required specialized expertise and careful attention to detail, including thorough surface preparation of the existing columns to ensure proper bonding and adhesion. Regular quality control checks were implemented throughout this phase to guarantee the efficacy of the FRP strengthening.

Improvements to the foundation's stability involved the installation of deep soil improvement techniques. Specifically, we employed vibro-compaction to densify the underlying soil and increase its bearing capacity. This crucial step addressed concerns about potential foundation settlement or lateral displacement during a seismic event. This phase involved extensive geotechnical investigation and meticulous monitoring of soil conditions to ensure the effectiveness and safety of the procedure. Prior to this stage, we conducted thorough dynamic analysis to validate that soil improvements would significantly improve the overall seismic performance of the structure.

Post-retrofit analysis, including both nonlinear static and dynamic analyses, demonstrated a significant increase in the building's seismic performance. The enhanced ductility and shear strength of the columns, combined with the improved foundation stability and removal of the brittle infill walls, resulted in a substantial increase in the building's ability to withstand strong ground motions. The results clearly exceeded the minimum safety requirements outlined in the current building codes and regulations.

The successful completion of this project highlights the effectiveness of a holistic retrofitting approach. By systematically addressing the building's various weaknesses, we achieved a substantial improvement in its overall seismic resilience. This project serves as a valuable case study demonstrating how a combination of different strengthening techniques can lead to significant safety improvements in existing structures, offering a reliable and effective solution for mitigating earthquake risks in older buildings. Furthermore, the project emphasized the importance of meticulous planning, careful execution, and rigorous quality control in all aspects of the retrofit process. The successful application of advanced materials like FRP further showcased the possibilities for modernizing existing infrastructure and enhancing its ability to withstand future seismic events. Careful consideration of potential disruptions to building operation, during the different phases of the retrofit, were meticulously planned and executed, demonstrating that effective seismic retrofitting can be achieved while minimizing operational downtime. The entire project successfully integrated the best practices of structural engineering, geotechnical engineering and construction management to deliver a safe and resilient building. This experience underscores the importance of investing in seismic upgrades for existing infrastructure as a crucial step in ensuring public safety. The comprehensive nature of the project and the thoroughness of its documentation served as an invaluable learning experience for the entire team involved, enriching the understanding and application of seismic retrofitting principles in the long term.

7 Future Trends & Research

Advanced simulation methods are pushing the boundaries of seismic analysis. Researchers are developing sophisticated computational models that can accurately predict the behavior of complex structures under various seismic scenarios. These models incorporate advanced material models, nonlinear behavior, and the interaction between soil and structure, leading to more reliable assessments of seismic vulnerability and the effectiveness of retrofitting strategies. This is particularly crucial for irregular or complex structures where simplified methods may fall short. The integration of machine learning techniques is also transforming the field, enabling faster and more efficient analysis of large datasets, optimizing the design process and leading to more cost-effective solutions. The accuracy of predictions will significantly influence the future of building codes and the design of safer structures. Furthermore, advancements in high-performance computing will allow for even more detailed simulations, improving the capacity to model complex scenarios such as liquefaction or near-fault ground motions.

The integration of smart technologies into seismic retrofitting is revolutionizing how we approach building safety. Smart sensors embedded within structures can provide real-time monitoring of the building's response to seismic events, identifying potential weaknesses and predicting imminent structural failure. This continuous monitoring allows for timely interventions and preventative maintenance, reducing the risk of catastrophic damage. Data collected by these sensors can also be used to refine existing models and design more effective retrofitting strategies, tailoring solutions to individual building characteristics. For instance, advanced sensor networks can detect subtle changes in structural integrity, alerting building managers to potential problems before they escalate into significant issues. Such systems enable predictive maintenance, enhancing the overall lifespan and safety of the structure. Moreover, the integration of data analytics and machine learning allows for the development of predictive models that can forecast the risk of future earthquakes and the potential damage they might cause to specific structures.

One key area of research focuses on developing self-healing materials and structures. These innovative materials possess the ability to repair themselves after experiencing damage, enhancing their longevity and resilience. Imagine concrete that can automatically repair micro-cracks, reducing the risk of larger cracks forming under seismic loading, ultimately extending the lifespan of the structure and minimizing the need for frequent and costly repairs. Such self-healing mechanisms can significantly improve the performance of retrofitted structures, potentially delaying or eliminating the need for major interventions. Research in this area is exploring various techniques, such as embedding bacteria or polymers within concrete matrices that can generate calcium carbonate to repair cracks. The potential applications are vast, promising to enhance the durability and resilience of buildings beyond current capabilities. This approach could significantly reduce maintenance costs and extend the lifespan of structures in seismically active regions.

Another promising area is the development of advanced damping systems. These systems are designed to dissipate seismic energy, reducing the forces exerted on the structure during an earthquake. Passive damping systems, such as viscous dampers or tuned mass dampers, are already employed in some retrofitting projects, but research continues to focus on developing more efficient and cost-effective systems. Active damping systems, which use actuators to actively control the structure's response, are also attracting increasing attention. These systems offer more precise control over structural movement, potentially reducing damage significantly. These technologies are particularly relevant for tall buildings and critical infrastructure, where minimizing movement is essential to preserving functionality and avoiding collapse. Advanced control algorithms are also being developed to optimize the performance of these systems, making them even more effective at reducing seismic forces and ensuring building safety.

The role of big data analytics in seismic retrofitting cannot be overstated. The vast amount of data collected from sensor networks, historical earthquake records, and structural analyses can be analyzed to identify patterns, predict future events, and optimize retrofitting strategies. Machine learning algorithms are being employed to analyze this data, leading to more accurate predictions of ground motion, building vulnerability, and the effectiveness of different retrofitting techniques. This data-driven approach enables a more targeted and efficient approach to seismic risk management, focusing resources on areas and structures that need them most. By leveraging the power of big data, we can move towards a more predictive and proactive approach to building safety, rather than relying solely on reactive measures after an earthquake. This comprehensive, data-driven approach promises significant advances in earthquake resilience and structural safety.

7.1 Advanced Simulation Methods

Advanced numerical modeling techniques are increasingly vital for evaluating the effectiveness of seismic retrofitting strategies. These methods allow engineers to simulate complex structural behavior under seismic loading, surpassing the limitations of simpler analytical approaches. Nonlinear finite element analysis (FEA), for instance, can accurately model the material nonlinearity, cracking, and yielding that occur during strong ground shaking, providing a more realistic assessment of structural performance than linear models. This sophisticated approach is particularly important for evaluating the effectiveness of retrofitting interventions on complex structures, offering insights into the behavior of the building under extreme conditions that physical testing can't easily replicate.

Beyond basic FEA, advanced techniques such as finite-discrete element methods (FDEM) allow for the explicit modelling of material fracture and fragmentation, particularly relevant in the assessment of brittle masonry structures undergoing retrofitting. These methods can simulate the progressive damage and failure of different components, capturing the complex interactions between strengthened and unstrengthened elements during an earthquake. We can therefore more effectively pinpoint vulnerable areas and optimize the placement of reinforcing elements. The level of detail achievable with FDEM offers a significantly more comprehensive understanding than traditional methods, leading to more resilient designs.

The incorporation of advanced constitutive models within these simulations is crucial for accurate predictions. Constitutive models define the relationship between stress and strain for different materials, accounting for material nonlinearity, plasticity, and damage accumulation. Sophisticated models consider the effects of cyclic loading, a key aspect of seismic events. Using advanced constitutive models allows simulation of the behaviour of materials such as concrete, steel, and FRP composites under cyclic loading conditions, thereby improving the accuracy of seismic assessments. Carefully calibrating these models using experimental data from material testing is essential for reliable simulations.

Probabilistic seismic demand analysis (PSDA) methods represent a paradigm shift in seismic engineering. Unlike deterministic approaches that assume a single earthquake scenario, PSDA considers the inherent uncertainty in ground motion characteristics and structural response. Through the incorporation of probabilistic models for ground motion and structural fragility, PSDA can provide a more realistic assessment of the seismic risk, considering a wide range of potential earthquake scenarios. This comprehensive evaluation, crucial for retrofitting projects, provides a statistically robust estimate of the probability of exceeding various performance levels. This probabilistic framework allows for a more informed decision-making process when deciding on retrofitting strategies.

Furthermore, integrating advanced simulations with machine learning techniques opens up exciting possibilities. Machine learning algorithms can be trained on large datasets of simulation results to build predictive models, allowing for efficient and accurate estimation of structural response under seismic loads without the need for computationally expensive simulations for every scenario. This approach allows engineers to rapidly assess various design options and retrofit strategies, accelerating the design process and optimizing resource allocation. Such techniques also allow for the creation of simpler, surrogate models, which retain the accuracy of the simulations but have reduced computational demands, enabling their use in larger optimization and parameter studies.

However, it's important to acknowledge that advanced simulation methods, while powerful, have limitations. Accurate simulations require detailed knowledge of the material properties, geometry, and boundary conditions of the structure. Data acquisition can be challenging and expensive, requiring careful planning and precise measurements. Furthermore, interpreting the results of these complex simulations requires expertise in numerical modeling and structural analysis. Careful validation of the simulation results against experimental data is crucial to ensure their reliability. The accuracy of any simulation depends heavily on the quality and quantity of input data.

Despite these limitations, advanced simulation methods are becoming increasingly essential tools for evaluating the effectiveness of seismic retrofitting. They provide insights into the complex behavior of structures under seismic loading, improving the resilience of our built environment. As computational power continues to increase, and simulation techniques continue to advance, these methods will play an even more critical role in ensuring the safety of buildings in seismically active regions. The future of seismic design and retrofitting relies on an integration of cutting-edge computational approaches and experimental evidence for maximum impact and reliability. Choosing the appropriate simulation method for a particular project will always require careful consideration of the structural complexity, available resources, and desired level of accuracy.

7.2 Smart Retrofitting Systems

Smart retrofitting systems represent a significant leap forward in enhancing seismic resilience. These systems go beyond traditional strengthening techniques by incorporating advanced sensors, actuators, and control algorithms to create dynamic responses to seismic events. Imagine a building equipped with a network of sensors constantly monitoring its structural health, detecting even subtle signs of distress before they escalate into major damage. This real-time monitoring allows for immediate interventions, preventing catastrophic failures and significantly reducing the risk of collapse.

This proactive approach is a substantial upgrade from the reactive measures typical of conventional retrofitting. Instead of merely strengthening structural elements, smart systems actively manage the building's response to ground shaking. For example, actuators, controlled by sophisticated algorithms, could adjust the stiffness or damping properties of specific structural components in response to detected seismic activity. This adaptive behavior allows for optimized energy dissipation and reduces the overall demand placed upon the structure during an earthquake. The level of precision achievable through smart retrofitting is extraordinary, allowing for tailored interventions that focus on the most vulnerable areas of the building. This not only improves overall safety but also potentially minimizes disruption during an earthquake, facilitating a faster return to normalcy after the event. Furthermore, the data gathered by the integrated sensor network is invaluable for researchers and engineers. By analyzing the building's response during actual seismic events, it is possible to further refine the design of these systems and build more resilient structures.

The integration of these technologies requires careful consideration. The design of smart retrofitting systems must be tailored to the specific characteristics of the building and its environment. The accuracy and reliability of sensors are paramount; their placement must be strategically determined to capture critical information about the structural behavior. Failure of any component in the system could lead to a false sense of security, and this will completely defeat the purpose of enhancing safety. Thus, redundancy and robustness in the system design are essential. Moreover, the power supply and data transmission systems must be reliable and robust enough to withstand the destructive forces of an earthquake. A failure during an event could mean the loss of the critical information during an event.

Costs associated with smart retrofitting are currently a barrier to widespread adoption. However, as technology advances and production scales, the cost is likely to reduce. It's worth noting that the long-term economic benefits of preventing catastrophic damage far outweigh the initial investment. By minimizing repair costs, reducing downtime, and safeguarding lives, smart retrofitting systems offer a compelling return on investment. The potential savings and the life preservation are significant factors.

One of the most promising aspects of smart retrofitting is the potential for improved post-earthquake functionality. Many traditional retrofitting methods focus primarily on preventing structural collapse. Smart systems, on the other hand, can also aim to maintain essential services, such as power and communication, thereby reducing the impact of a seismic event on the occupants. Imagine a hospital or emergency response center equipped with a smart retrofitting system. Its continued functionality during an earthquake could be life-saving. This improved post-earthquake performance offers immense benefits in terms of life safety and societal resilience. The potential for saving lives during emergency events with this technological innovation is immeasurable.

Looking forward, research efforts should focus on developing more affordable and reliable sensors and actuators, improving the predictive capabilities of the control algorithms, and addressing the integration challenges involved in deploying such systems in existing buildings. Furthermore, building codes and standards need to adapt to accommodate the use of smart retrofitting systems, providing clear guidelines for design, installation, and maintenance. The integration of smart technologies into seismic retrofitting is not simply an incremental improvement but a paradigm shift. We are moving from passive strengthening to active protection. This shift holds the key to building a more resilient future. The future of safety in buildings relies on advanced and safe technology. We must embrace that.

www.ingramcontent.com/pod-product-compliance
Lightning Source LLC
LaVergne TN
LVHW022013200926
848481LV00007B/2369
* 9 7 9 8 3 4 9 3 9 0 2 9 6 *